GIFT
OF
CHRISTOPHER BROOKHOUSE

The American Poetry Series

VOLUME 1 *Radiation*, Sandra McPherson

VOLUME 2 *Boxcars*, David Young

VOLUME 3 *Time*, Al Lee

VOLUME 4 *The Breathers*, James Reiss

VOLUME 5 *The House on Marshland*, Louise Glück

VOLUME 6 *Making It Simple*, David McElroy

VOLUME 7 *Goodwill, Inc.*, Dennis Schmitz

VOLUME 8 *The Double Dream of Spring*, John Ashbery

VOLUME 9 *The Collector of Cold Weather*, Lawrence Raab

The Collector of Cold Weather

Lawrence Raab

The Collector of Cold Weather

The Ecco Press New York

First published by The Ecco Press in 1976
1 West 30th Street, New York, N.Y. 10001

Published simultaneously in Canada by
The Macmillan Company of Canada Limited

Printed in the United States of America
The Ecco Press logo by Ahmed Yacoubi

Library of Congress Cataloging in Publication Data
Raab, Lawrence, 1946-
The collector of cold weather.
(The American poetry series; v. 9)
I. Title.
PS3568.A2C6 811'.5'4 76-3301
ISBN 0-912-94632-6

Grateful acknowledgment is made to the following magazines in which these poems first appeared: *The American Scholar*: "The Assassin's Fatal Error." *Antaeus*: "The Collector of Cold Weather," "Riddle," "Pastoral." *Berkshire Review*: "Attack of the Crab Monsters." *Marilyn*: "Doctor Watson's Final Case," "Accidents of the Air." *The New Yorker*: "Visiting the Oracle," "Summer Poems." *The Paris Review*: "Further Adventures of the Pipe." *The Poetry Miscellany*: "Each Time You Carry Me This Way," "The Poem on Your Table," "The Invisible Object," and "The Island of Lost Souls." *Prairie Schooner*: "The Blue Histories of the Wind." *Shenandoah*: "Water."

I would like to thank Syracuse University,
the National Endowment for the Arts,
and the University of Michigan Society of Fellows
for their generous support.

—L.R.

For my mother and father

CONTENTS

1

With habit and repetition he gained to an
extraordinary degree the power to penetrate
the dusk of distances and the darkness of
corners, to resolve back into their innocence
the treacheries of uncertain light,
the evil-looking forms taken in the gloom
by mere shadows, by accidents of the air,
by shifting effects of perspective. . . .

—HENRY JAMES
"The Jolly Corner"

PELLÉAS You're not thinking of me right now.
MÉLISANDE But yes, I only think about you.
PELLÉAS You were looking somewhere else.
MÉLISANDE I saw you somewhere else.

—MAETERLINCK
Pelléas et Mélisande

PASTORAL

Today in Peru, this first day of summer,
how easily another life
falls into your hands—marriage,
dreams, tables and chairs, and the same
incurable disease. Here
in this light you're
the handsome but dishonest
jungle doctor, happy to help the old man
to his grave, happy to find the daughter
beautiful, despondent, and anxious
for medication. *Today* you write
in your journal *in this helpless country*
I must pretend to overlook
the question of the inheritance. Likewise tomorrow
you will choose to ignore
reports of an army of ants
crossing the border into the nation.

Like the rest of us, in the other countries,
you've grown unwilling to accept
the madness of evidence, and so prefer
to walk until morning
beside the cold beds of night-blooming flowers,
considering the injuries of fog,
the small advances
of moonlight. Is it possible,
you wonder, she has not yet
noticed my affliction?

Evenings in Peru
twelve centuries of hoarded darkness
glitter in the faces
of the jewels she kept for herself.
Even the day of the wedding she would not
trust you. Nor would the servants
ever be your friends. *From that first afternoon*
the pulse of light in the foliage seemed
unnatural, and the thin music
of the jungle became too quickly the sound of trees
collapsing at a great distance.

No one is content,
in the country, in the middle of summer.
For weeks I have dreamt of waking up
on the table in the white surgery.
Back home in the winter
you'd tell your friends how the new life was like
the story of the doctor in Paris who knew
his work would never carry him
far enough and who said
at the end, "There's something here, but I can't
tell you what it is."

Once she had played for hours
Mozart on the harpsichord. It was the night
she left for good, taking the little purse
of diamonds, and all the silver—
forks and spoons—while you thought
the music was meant to say
everything had been revealed
when you were not watching. And you wrote:

But if she was that beautiful
it's no excuse, it never was.

Your plan remained
the same: leave the explanation
to the end or leave it out
altogether. Still, you could not help
but hear, each night, the muffled thud of your heart,
and of hers,
and the sound also of teeth
in the boards on the windows. Now

your house fills
with the paraphernalia of summer, as
the black swarm closes in—called
"the descent of spirits"—and after it
the famous yellow light
of autumn in Peru
drifts through the immaculate
bones of your fingers
and settles around you like
the finest gold of another distant country,
or a different season, or another life.

RIDDLE

That moon, for example—
shaving from a nail,
palest of openings.
And those stars—
they're fading now
but they were always fading.

From evening I take
what I need to get through
the daze of morning and afternoon.

Water—for the sake of company.
Fire on the hill.

Each night I show you this
white cone of ash. Each night I tell you:

But I keep nothing for myself.

Hedge and field,
curve of the road.
Table and book,
coffin in the ground.

Whatever settles
into the earth

comes back to the light
in time.

Again, I tell you this
and you believe it.
And you take what you need,
and you believe it is yours.

WATER

I

Whichever way water
turns it touches
itself turning in another direction

Invisible now
now reflecting whoever
finds himself looking
beneath the line of the wind

You remember the rules

Water seeks the level that pleases it
making a place for itself
wherever it chooses

calling everything
it touches its own
and falling back
in its own good time

2

Hundreds of feet beneath you
it creeps along a fault
drop by drop widening the rock
softening an edge
breaking off a splinter

So a cave blossoms

Water counts the time but does not care
You could learn from it
Speak to it of your troubles
Ask about your wound why it
refuses to heal

Ask about absence

Water has spent a long time learning
how to fill with itself
the space of a missing thing

3

Wherever it can go water goes

On your window
the early frost has drawn a map
and the small cloud of your breath
fades from the blade of the knife

The shape of someone like yourself
drifts in the shelter of still water
You reach down

A maze of circles meets your hand

EACH TIME YOU CARRY ME THIS WAY

From the paintings of Morris Graves

The masked bird fishes in a golden stream

Flight of plover
loon on an autumn lake
moor swan   wounded gull
moon mad crow in the surf

The little-known bird of the inner eye
sends this red message

Consider the eyes
looking at you
Consider the space
gathered into the shape of home
and these beginnings
tokens of the journey—

purification in the stone
darkness in the chair
old ground wild with a sudden light

Bird in the mist
spirit bird   eagle in the rock
shore bird submerged in moonlight
guardian with horns

When the eye falls back to stillness
sea   fish   and the silence of stars
blossom on the surface of inner walls

The blind one
tightens his claws around the stone

and the dark-eyed bird
listens
for the reply of the silver minnow in his beak

What direction can I give
when you ask the same questions
each time you carry me this way

SUMMER POEMS

For my grandmother

This rain will stay with us all night

It has no story of its own
but we can make one up

and the sleep of rain will be
the sound of pages turning
here in the dark grass

Which of us
left that book out on the lawn
and who was the child then

Whenever I hear thunder
I go to the window

❧

In the photograph of water the sky is lost

We sit together in the pale field
the dog at your feet
myself on your knees
listening to a story

Crescent Beach   Shaker Bridge   Mount Calm

The rest of the world
starts just across the lake
Crows perch on the telephone wires
Fat cows stand knee-deep
in their own slow afternoons

and already I have forgotten
everything you were saying

From the porch you smile up at me

I find my bed at the top of the stairs
The trees gather at my window

Mum   I never expected them
not to change

At night the tiny lights of cars
cross Shaker Bridge
and climb the mountain

Where could everyone be going

Theirs was the story
I would have chosen as my own

Later
wind shakes the trees
and together they become
the sound of water

as the heavy quilt lifts
from me

and far away

the rain begins

THE POEM ON YOUR TABLE

For Robert Pack

Inside the poem
a girl walks down a street
A man composes a letter
The first hush of winter divides them

At the end of the poem—
a long white silence
and at the beginning
the same
Around the poem the real world gathers
where just now snow descends
making so soft a sound you hear
your heart keeping track of the time

Again you hold the poem in your hand
Again inside it
everything that has already happened
begins

The terrible white page stretches
in front of the man sitting at the desk
The girl gathers her things together
opens a door
and walks into the dark where
the first snow of the year
falls around her
and makes of silence
a voice she can almost hear

She does not know where she is going
or what she may have left behind or lost
As she listens to her footsteps
marking the time
she does not remember how
she found herself inside this poem

THE INVISIBLE OBJECT

Held toward light
in the shape of your hands
it's clearer than water,
clearer than glass,
than air.
It holds nothing back.

Against the sky, sometimes
it's red, sometimes it's blue.
Set in the grass in your garden
the bees stumble through it.

Looking closely
inside there's always another day
where your life could be.
Yet it has nothing to do
with dreams that change
when you change your mind.

On a table at night it becomes
the center of your house,
a small fire in a darkened room,
and the silence bends toward it
and touches it with one finger.

Lost for days it appears
suddenly
in a pile of old hats.

No reason to be there,
no reason to be anywhere else.

FURTHER ADVENTURES OF THE PIPE

Homage to Magritte

Sitting down, unfolding
the paper, without looking
you raise the light toward the pipe.
But it's not a pipe, it's a pear.

Surely it was a pipe
only seconds ago!
What might account for this
unusual transformation?
Could it be a sign
of something important?
Can the pear be eaten?

No. The pear's as hard as stone.
In fact it is a stone,
disguised as a pear.

Perhaps the pipe was stolen—
unlikely but possible.
Perhaps the stone pear
can be put to use
as a paperweight, although
you have no need of one,
or a doorstopper, or some kind
of decorative object.

Then you notice the inscription:
This is not a pear. It's a pipe.

And everything is clear.

ACCIDENTS OF THE AIR

Like nothing in this world.

Not the sound of the shirts
being sewn, the ladders
lifted, the water poured,
or the tree
filled with wind.

Another story, then. Another life.

Where trees become
more like single leaves, more like
the legs of tables, and you
discover the reports
of witnesses.

They say *It's not the chipped*
skeletons in their cages
or the cold
arrival of the lists.
And it's not the silence.

But another life
altogether.

Where light
at evening crosses
the lawn
like nothing you had imagined

in this life or any other.

2

Sleep is a condition in which I refuse to have
anything to do with the outer world and have
withdrawn my interest from it. . . . Our relationship
with the world which we entered so unwillingly
seems to be endurable only with intermission. . . .
It looks as if we do not belong wholly
to the world, but only by two-thirds;
one third of us has never yet been born at all.

—SIGMUND FREUD
Introductory Lectures

Yet a little sleep, a little slumber, a little folding
of the hands to sleep;
So shall thy poverty come as one that travelleth,
and thy want as an armed man.

—*Proverbs* 6:10–11

THE BLUE HISTORIES OF THE WIND

I

Again, the ominous sounds of furniture!

The small hooked rug edging toward the door.
Chairs rubbing softly against the desk.
Books slipping from their places.

By morning two more pictures
have disappeared into the wall.
I discover faint traces: a landscape
with cows, a stream, a mountain;
an old man with a beard.

Out on the lawn I complain to the manager.

"The view is not what I was promised.
Something is wrong with these books.
And where has my room gone?"

"Yes, yes," he replies, "everything
has been taken care of."

2

Often, I unlock
the tiny photograph you gave me.
It grows darker
each time the sun touches it.

My work goes badly here.
I keep losing track. Every day
things retreat a little further,
and now even the subject escapes me.
What is this which blocks us?
Mirrors? Sun spots? Or just ordinary weather?

Again I hear your voice
returning from some unlikely distance,
broken into by the blue wind,
or today's list of excuses, or another chair
crashing through the windows of the dining room.

Then everything begins to be the same.

You drift further away
and when you return
you are changed, and then I
too am changed.

3

More than I could have expected
has been removed.
Even the plastic geraniums have disappeared,
surely no loss,
though I had grown used to their
pure, unalterable colors.

About the city I can add
little to the other letters.

Today it seems
I have business there.
Addresses and appointments appear
next to my napkin at breakfast,
but if I delay long enough
they are taken away.

Have I told you about the others?

M., who spends all day drawing maps,
and R., who carries a gun and will
never speak to me. Nor can I discover
what they want.

No one goes to the beach any more.
The bright umbrellas are gone.
The ocean grows colder and more silent.

Tonight snow is predicted.

What else can I tell you?

4

"No, not *dead*, Doctor. That body
has never been alive!"

Little flames jumped from the electrodes.
He began waving his hands and shrieking:
"This monstrous experiment of yours
challenges the very order of creation!"

"Yes! Electricity is life!
I imagine an entire race of men
whose only wants are electricity!"
Lightning snapped through the machinery.

"A hundred thousand megavolts, if necessary!"

Then I remembered the end of that story—
villagers advancing into the forest
with their torches and pitchforks, and the monster
stumbling through the mire toward the windmill,
or the quicksand, or back
to the black pits beneath the lab.

"But in whose darkness, Doctor, will we meet?"

Weeping a single tear, he pulled the fatal lever.
"We belong dead."

5

It's like that.
Hard to get in, and hard
to get out again.

Weeks pass, and months,
or so it seems. Years.

From time to time
I hear footsteps on the stairs,
coming up, going down.

Then quite suddenly
after a night of storms
it begins.
I pick out a description,
explanations, a middle
and one possible end.

The intricate machinery of the rain!
The clever traps of the natives!
The erratic, swooping flight of bats!
Everything fits.

And finally a knock on the door.
"I saw the light so I came."

It doesn't matter, I reply.
Nothing matters except
that you've come back.

But of this invention, I note,
I cannot be certain.

Tomorrow all these pages will be empty.

6

I count the choices that remain,
all preparations having deserted me.
I continue to collect whatever may be needed,
but things will not stay put.

Snow is predicted.
M. tells me, "When the snow arrives
it pretends to be everywhere,
but no one believes it."

Even my own notes baffle me:
"I must have the diamonds!" Or:
"Consider: the blue histories of the wind."

"When the monster saw the first torch,"
M. says, "no part of him
wanted life at that price.
The patched-up heart pumps on and on.
But part of us was never born."

When I took out your picture
you were all but gone,
as if you had not been there,
as if suddenly
I could not be here.

7

Huge snowmen have appeared in the public parks.

Last night I found myself
walking through the ruins of the snow.
I had given up the laboratory,
dismantled the machines,
buried the notebooks.

Though in each was some part
I could be sure of.

But what could I ask for?
I never wanted to live here.

If only I could imagine another country!

If only I could think of nothing
but the perfect lives of strangers.

8

When my own voice found me
it had mislaid something important.
What was it? When
did you have it last? Remember!
But it will not remember.

On my table: the shape of a window
with a blue jar and a single cloud.
Invisible objects.

Even the weather possesses its spells,
as sleep its divisions and one
constant deception.

And these messages—
to what conclusion?
You are left, you always are left.
At the end of my words
you cannot answer back.

I should have known.
I should have expected it.

9

The tall man who travels,
the armed man who never speaks,
my companions.

Our journeys connect
in this hotel at the edge of water,
where whatever we can discover
remains
hidden by what it shows.

Together we walk out into the winter.
Its light turns on us
with a hundred edges.

10

Unfamiliar territory opens around us.
We watch our steps.

We pass the snowmen with their blind, smashed faces.
We pass the Doctor. "What is one life,"
he says, "in the service of those mysteries?
Look at the heart, the heart!"
We pass the old man with the beard.
He says he was your father.
He says, "I am blameless. Let me pass."

The tall companion unfolds his maps
while snow attempts to circle us
with its pale confusions.
But that's an old trick
and I know the answer.

Look. When I unfold my hands
they fill with intricate disguises,
all that I have managed to save, all
that I have invented to save us.

This way, the blue wind replies.

The storm divides before us,
closes after us.

This way, this way.

3

Among these unfinished tales is that of
Mr. James Phillimore, who, stepping back into
his own house to get his umbrella, was never
more seen in this world. No less remarkable
is that of the cutter *Alicia*, which sailed
one spring morning into a small patch of mist
from where she never again emerged, nor was
anything further ever heard of herself and
her crew. A third case worthy of note is that
of Isadora Persano, the well-known journalist
and duellist, who was found stark staring mad
with a match box in front of him which contained
a remarkable worm said to be unknown to science.

—ARTHUR CONAN DOYLE
"The Problem of Thor Bridge"

Music, states of happiness, mythology, faces
scored by time, certain twilights, certain
places, all want to tell us something, or
told us something we should not have missed,
or are about to tell us something.
This imminence of a revelation that does not
take place is, perhaps, the esthetic fact.

—JORGE LUIS BORGES
"The Wall and the Books"

VISITING THE ORACLE

It's dark on purpose
so just listen.

Maybe I inhabit a jar, maybe a pot,
maybe nothing. Only this
loose end of a voice
rising to meet you.
It sounds like water.
Don't think about that.

Let your servants climb back down the mountain
by themselves. I'll listen.
I'll tell you everything
I discover, but I can't
say what it means.

Someone will always
assure you of the best of fortunes,
but you know better.

And keep this in mind: The answer
reveals itself in time
like the clue that fits
perfectly and explains everything
after the crime has been solved.

Then you will say: *I should have known.*
It was there all along
and never even concealed,

like the story of the letter
overlooked by the thief because
it had not been hidden.
That's the trick, of course.

You don't need me.

ATTACK OF THE CRAB MONSTERS

Even from the beach I could sense it—
lack of welcome, lack of abiding life,
like something in the air, a certain
lack of sound. Yesterday
there was a mountain out there.
Now it's gone. And look

at this radio, each tube neatly
sliced in half. Blow the place up!
That was my advice.
But after the storm and the earthquake,
after the tactic of the exploding plane
and the strategy of the sinking boat, it looked

like fate and I wanted to say, "Don't you see?
So what if you're a famous biochemist!
Lost with all hands is an old story."
Sure, we're on the edge
of an important breakthrough, everyone
hearing voices, everyone falling

into caves, and you're out
wandering through the jungle
in the middle of the night in your negligée.
Yes, we're way out there
on the edge of science, while the rest
of the island continues to disappear until

nothing's left except this
cliff in the middle of the ocean,
and you, in your bathing suit,
crouched behind the scuba tanks.
I'd like to tell you
not to be afraid, but I've lost

my voice. I'm not used to all these
legs, these claws, these feelers.
It's the old story, predictable
as fallout—the re-arrangement of molecules.
And everyone is surprised
and no one understands

why each man tries to kill
the thing he loves, when the change
comes over him. So now you know
what I never found the time to say.
Sweetheart, put down your flamethrower.
You know I always loved you.

THE ISLAND OF LOST SOULS

Perfectly cold and blue as the light
fed through the wire, the orchid
opened in my hand. Glass clouded
with its heavy breath, petals
twisted from the stem

and broke. And the dead eye of the crystal
stared back at me.
Because he was mine the thing I'd made
screamed when he felt the light. And would not
speak. Because he remembered me.

But I was telling you how I took the orchid
and stripped away a thousand years
of slow evolution
to prove the secret
lay in the surface

where no one thought to touch it.
But I can see the question that divides us
is this question of the pain.
In all that we do
it drifts back, covering

our scars with its skin,
and the stubborn flesh
grows, day by day,

to defeat us—an hour lost and then
one year and then ten.

Light was a mirror, inside the glass.
I could have stepped through
and no one would have stopped me.
But I heard a voice and I went on
the way it led me, and it led me here.

The orchid unfolded in my hand. Here,
you can see its mark. And here—
where I twisted the heart when I thought
I could burn the pain away,
when I thought it wasn't mine.

DOCTOR WATSON'S FINAL CASE

For Jonathan Aaron

Regarding these most
recent outrages we sincerely hope
there will be no repetition.
Enough damage has been done
although, I must admit, we did not
lack advice, or warning—
twelve, in all, anonymous letters.

Don't you realize there's
a monster at large
in this city
bent on destruction?

But no one knew how long
you'd been nosing around the waterfront
disguised as a gorilla, the same gorilla
found missing from the zoo exactly
one week to the hour before
the night of the crime. "I don't know
where he could have gone," the keeper told
the reporter, his eyes filling up
with tears. "I always thought
he was happy here."

"This man," you said,
catching the orderly by his sleeve,
"died precisely
at the stroke of twelve. Observe

the ash from his cigarette." Naturally
a full confession followed.

"Well, sir, it was all my fault.
I know I should have spoken out
the first night they arrived at the office.
But the sight of blood oozing beneath the door
seemed to awaken murderous instincts.
Then Brezard was gone.
'Are you going to let this happen?'
they said to me.
'Are you going to throw away
the chance of a lifetime?'
What choice did I have
after all?"

"It's reasonable," you concluded,
"because it's true."
Later you explained:
"If a man is ugly, he does ugly things."

Thus we find no necessity
to release at this time
the whole story concerning
the politician, the lighthouse,
and the trained cormorant. I maintain
no one would have understood
in any case.

"On the contrary," you reply.
"There is at least one reader
who would have understood.
There is always one reader
who will understand."

THE ASSASSIN'S FATAL ERROR

When in doubt have a man come through a door with a gun in his hand.—RAYMOND CHANDLER

He comes through the door,
the big gun in his fist. He says
"Nobody's going anywhere."
Nobody was, nobody's
even here, except
me and this bottle of scotch,
and I'm used to waiting.

He tells me to explain
about the pearls because he knows
ways to make me tell. However,
I know nothing about them,
nothing about Mr. R. or The Big Man,
and nothing about Oregon where I
have never lived at any time in my life.

Perhaps I'm lying, but he's convinced,
although he will shoot me anyway,
which I understand.

Perhaps I say, "What kept you?"
Probably I just finish the scotch,
which is third-rate but effective.

Mine, you understand, has been
a temporary disguise, which may or may not
be explained at a later time.
Its importance to the story

lies in the discovery of the body
by the detective, tomorrow.

I also turn up three chapters from the end
as a Doctor, where I can be trusted
even less than now, when I still have
this death to get through.

"There must be connections,"
I tell him. "There always are.
And it's smart to leave
the witness silent, get the job over
and get out of town."

The gun wanders around the room.
"Listen," I tell him. "Anything
can happen. But this could be
the fatal error. You don't know
any more than I do."

The long tube of the silencer turns toward me.
I consider the finger on the trigger,
the sound like somebody coughing
upstairs in an old building,
and now the single bullet
suspended in the air between us.

You could ask: But what did I expect?
And I would have to say: Only this.
Nothing but this.

THE CONFESSIONS OF DOCTOR X

The flames will never reach us here.
That's what I said.
I knew the laws were wrong but I couldn't
prove it. I couldn't stop
and I couldn't go back,
for the dead travel fast
and you were so beautiful
when you were alive.

I'd wasted years of explanation.
"Fine operation or not,"
the Professors told me, "some things
are better left alone."
And when complaints came in
about the digging up
of graves, or the occasional
unfortunate death, "Something," I said,
"Something must have come down
out of the hills.
Wolves, maybe."

I discarded a little more
each day, everything I could manage
to forget, while the stillness
of a single room became
the equation that would not balance.

Like the threats of glass.
Like the wires, piercing everything

I had assembled.
Or the little fires that moved
from room to room
when I wasn't watching.

Like the only dream you told me:
None of this has to be true.

The piano. The green velvet couch.
Your coat tossed over
the arm of a chair, as you left it
that last afternoon before
the accident.
Paintings darkening
on the walls. White
sheets of music. Light
like sunlight.

Mademoiselle, the fire of the sunset
in your eyes
consumed my heart!

My solutions left nothing
to chance. I said
the flames would never reach us here.
I said we could start over.
Until the generator choked and froze
and the walls gave in, I was certain
you would understand.

But only the body lives
after death, and the heart
is lost
in the blind calculations of the blood.

4

THE COLLECTOR OF COLD WEATHER

I THE LOST NOTEBOOKS

Winter cut across our tracks.
Daylight caught us in our sleep.

The only dream I could save
was the one the Doctors sold me—
twelve windows and twelve doorways,
the flask of ice
in the shell of ruin,
and the glass star
inside the cage.

I'd told them all before,
"This isn't the life I thought was mine.
Is that where I was yesterday?
That sky, those mountains?
I'd never have known."

Heading south from the polar regions
everything was clear,
like the names we offered
to lakes and mountains,
like the intentions of the assassins
trailing us with their shining machines,
their shuttered coaches
never less than a day behind.

I knew the explanation

would be theirs in the end—
the long story of the buried light
of the last great stones.

2 THREATENING WEATHER

In the Black Museum
the Doctors invented a formula
for the solution of sleep.
I read their advertisement:

   It Pays to Cheat Death
      Fantasies Fulfilled
      Dreams Rewarded

Weeks after the bearers
abandoned us, we were reminded
of former lives,
the sweet predictable adventure
plotted out to the end.
It might be this one, we thought.
And then again it might not.

Snow doubled back without warning.
The sparkling rings twisted from our fingers.

3 THE WITNESS

But I was looking the other way.
Why should it always happen
so quickly and somewhere else,
the visibility bad, the food indifferent,
and your binoculars fogged by clouds
pushed in from the sea?
A voice is lost in that haze.
A face is pulled away. No one
knows how long.
On the table in the old hotel I noted:
the gold-rimmed glasses,
the walking stick capped in silver,
and two revolvers in a velvet case.
They knew what was expected.

4 THE CAVE OF WINTER

Who now remembers
Edward and Esmond and the good
Sir Alexander, whose box of cigars
was discovered six hundred miles away
in the Yellow Sea? Who can recall
those extravagant figures—
Doctor Tulip and his Magic Box,
Brother Amber, or the Puzzle King,
whom the gypsies feared?

Their ghosts
stumble down the paths of misfortune
like forgotten tribes of the Andes.
They press their cold hands against
the sheeted windows of the houses
of the recently dead.
Sons? Ancestors?

Their words are the same as yours
when you least expect it.

Not now, they say. *Later.*

5 THE HANGED MAN

It was the afternoon of the piano,
the evening of the guitar.

All night through the polished bars
I could hear the hammering—
the white scaffold
at the end of the street.
Air opened. Pigeons
blundered among the struts,
and the looped rope whistled to itself,
as you said it would.

Then it was
the morning of the glass harmonica,
that thin, high music beloved
of only a few, and understood by no one.

6 IN THE PALACE OF THE PUZZLE KING

We all felt uneasy
staring at the endless yellow
beds of chrysanthemums,
complimenting the monks on the perfect
placement of the stones, drinking
the king's pale tea.

I could only think
of that one empty square, the little train
on the horizon, the shadows
coming in fast on everything,
climbing walls, covering the chipped stonework,
and windows with the sky pasted in them
and perhaps a single passing cloud.

But I didn't want to be there.

Edward complained. "It will take hours
just to get out of this maze."
And the enemies would not be sleeping.
We could be sure of that.

They think nothing escapes them
but they're wrong.

The machinery of the mirrors
confounded their plans,
and the sorrows of the weather
drove them from the balcony.

7 A BOOK OF PICTURES

Not snow then but the deceptions of dust,
and avenues of the palest smoke leading
into nameless crippled towns so far off
who'd want to live there?
Who did? We moved on, we came
to an ocean on one side, we remembered
an ocean on the other. Turning back
we found our own roads going
from here to there and never stopping.
That was the end,
one of many that would follow.

Out of the cave of winter
I came back
just to make sure.

You don't have to think, I said.
You don't have to do anything.
You can sit with the gun on your knees.
You can remember whatever you want,
and if you feel like shooting
you can pull the trigger any time.

Talking about Sheriff Jim.
Talking about Boneyard Pete
and sweet Johnny
cut down at seventeen,
killed twenty men and fell over
with both hands in his pockets.

8 SORROWS OF THE WEATHER

The eye opens
but does not obey.
So you're deceived a second time
and a third. What you could not see
remains,
humming quietly to itself.

One night
becomes another. It's not what you were
looking for, but it's just as good.

Try to remember
the blessings in disguise:
the burned-out gallery,
the washer who sits with herself in the twilight,
the husks of the hearts of those whose feet
never touched the earth,
the water-horse, and others.

Light without shadow, passing
always to the west, away from us, as far
as anyone could hope to travel.

9 FAMILIAR OBJECTS

We were reading the diaries a second time

when last year's music rushed
through the house,
pulling the plaster down, destroying
the furniture.
"Wait a minute," I said.
"We might need this room again."

Too late!

"There must be some mistake."
"I'm not who you think I am."

"What about the hanged man?" I asked.

"Reference to the story of the hanged man
does not apply in this case."

Who would have suspected?

10 THE HOUR OF TRACES

For an hour everything is green.
Then blue. Then white.
I walk through it when it's green.
Leaves. Grass. Still water.
It's the best story of all.

For example: I'm on my way
to the palace, carrying
urgent reports for the king.

Help by Friday or all lost.
Enemy approaching. Supplies gone.
Send help. Send helicopters.

I say, "This is the life."
The road is green. The bark on the trees.
The twittering of anonymous birds.
And then it's blue.
A circle of fire, noiseless, falls from the sky.

I say, "I have my instructions."

11 THE APPARITION IN THE COURTYARD

After the examiners had departed
and the investigators had returned to the inn,
a few messages appeared on the wall.
"No, no," said Edward. "That's not it.
Let's try again."

Whose walking stick?
Whose glasses?
Whose gun?

Edward, I could hear you calling.
Esmond! Billy!
I could hear all of you
down by the beach in the fog.

12 IN THE BLACK MUSEUM

The stories of the others
continue to intrude. Certainly
they meant too many
different things.

What can be added to such unfavorable signs
and what can be covered that hasn't
already been in a more appropriate
time and place?

Try counting. Repeat
the numbers you know.
Or the consolation of, for instance,
sheep or ducks.

And this also may be useful:
A string with nine knots
concealed for nine days upon an enemy
will destroy him.

The problem, of course, is to get
close enough.

13 THE GYPSY'S CHARM

She said: "Of the colors we have
to offer only white is left.
You can have it cheap today.
And a polished glass.
And a printed explanation."

Sometimes I'd turn around
and discover a stranger hunched
among the pine trees, howling
and scraping at the loose bark.
"Look here," I'd say. "Is that your tree?"
Now I know better.

Keep the arm steady.
Sight along the barrel,
both eyes open.
If you see anything move
don't ask questions.

She said: "But at the end
it's cold, at heart
it's colder than you can imagine.
Remember that.
Keep that in mind."

14 THE MACHINERY OF MIRRORS

"I merely record the fact that I saw
something blacker than the trees
pass along the path toward the lake."

Followed by a ringing of bells
and the smashing of the china.
In the Blue Room a single table lay
on its side.
"It moves!"
"It rises!"
(The villagers claim to see a light,
but can they be trusted?)

Considering the lateness
of the hour, all was not
as it should have been.
(The villagers will not cross this spot after dark.)

The investigators noted:
1) The disappearance of the keys,
2) The incident of the jumping soap,
3) The curious occurrence in the cat's cemetery.

15 THE GAME OF THE GREEN MAZE

Each time the story is repeated
it's shorter.
Each time the end is different.

Advisors pluck at my sleeves
anxious to deliver the news,
the reports of distress from the city.
"Wait," I say,
"One or two important things
have not yet happened."

Sleep falls through the branches.
Not even the wind escapes it.

And before you know it you've
turned a corner and you're somewhere
else, opening
a box, for example, taking out
a small gun,
although the snow
has not yet ceased to fall, although
the slides and rules of the carpenter
remain on the unfinished table, by the statue,
in the center of the square, to the left
of the train, not far from the scaffold
which creaks in the least wind
at twilight,
the hour between the dog and the wolf.

16 THE PALACE OF WINDOWS

After the price of everything,
after the great cost, only you remain,
nameless,
like an invisible ray
concealed among the debatable intentions
of the shrubbery.

Who could have been calling—

"Here I am! Here I am!
When will you pay me?
When will that be?"

Not now, I said. Later.

Rain in the windows, all afternoon,
all night. I told you,
"This isn't the life I'd planned on.
What happened?"
"Three gray geese in a green field grazing.
I always remember that," you said.
"I used to love to see it."

Deaths we had foreseen, and the walkers
who followed us

with their unspoken conclusions,
their numbered pages,
their unfailing sense of direction.

"Gray were the geese and green the grazing."

17 THE SOLUTION OF SLEEP

Now you lean over your book,
your face falling among the paneled hallways.
Five hundred pages of rooms!
But it's late. It's winter.
It's evening in all the windows.
As the fire burns
the pink roses on the wallpaper
shudder and close.
The piano plays an old song
no one remembers. Then
it's too late. The roses
have moved into the windows,
into the tall mirror guarding the desk,
into the white notes of the music,
into the book. Your hand
slips from its place.

18 THE GLASS STAR

Cold settles into my fingers.
I tell myself—If I refuse to believe,
nothing can harm me. At the last
possible second I will slip
the noose from my neck and fall
through the blue air
with the rope, the perfect sign of my escape,
swinging out behind me.

Words will awaken in the pages of my sleep
between midnight and morning
in the submerged rooms of the city.

And your silence will move beside me
just for a moment.

Lady, tell me what I owe.
I know you never forget.

19 NIGHT WORKS

The collectors of cold have assembled
their evidence, and the professors
have completed their book.

Yes, I have invented my life
down to the smallest detail.
It follows me, listening and taking notes.
It enters a plea of innocence.
"Your honors, he really didn't know
what he was doing."

Left alone I would have been the owl
with a face like a white heart,
living among ruins
and the bones of my enemies, rising
into my stillness at evening,
inevitable, without mercy.

The judges stare sadly at my unfolded hands.
I have nothing left to offer.
I can hear the snapping
of bones, farther away than I ever
thought possible.

I'd like to sleep
and wake up about twenty
years ago. That's
my only request. To look out

from the tall upstairs window as if
I had something important to say,
some last word to clear up everything.
And then I'd
just let go.

But the investigators were deceived,
the assassins foiled, the Doctors astonished.
I merely record the fact.
Last night for the longest time the figure
of our friend leaned from that window,
gazing down at the ruined lawn as if
waiting, as if, I thought,
about to speak. Then he was gone.

"Nevertheless," they said, "the case
is closed. No one would believe you
in a hundred years."

Inside, pictures slid from the walls.
Snow pressed through the splintered doors.
Snow gathered in the gallery,
in the upstairs hall, along the eaves of the attic.
Weeping could be heard
all over the house.
Cries of impatience.
Applause and congratulations.

"What realistic tears!"
"What sincerity!"

Angels,
get my mansion ready,
for we are crossing the misty river,
one by one.

21 THE LAST MINUTE

Wind arrives when it is needed,
picking up the torn end of another year,
keeping its promise.
Light falls into place
among the difficult stones,
beside the buried story
we did not discover.

At the last minute someone enters
with a gun in his hand.
"No," I say.
"There's a time and a place for everything.
Go back."

I kept falling.

I passed the doors.
I passed the swinging rope.
I passed the empty rooms,
one after another, windows smashed,
floors littered with the toys of the rain.
I kept on falling.

I heard the water's dark refrain.
I heard the clock's one word.
I kept on.

A NOTE ABOUT THE AUTHOR

Lawrence Raab was born in Pittsfield, Massachusetts, in 1946. He received a B.A. from Middlebury College and an M.A. from Syracuse University, where he won the Academy of American Poets' Prize in 1972. His awards include a Book-of-the-Month Club Fellowship, a grant from the National Endowment for the Arts, and the Robert Frost Fellowship from the Bread Loaf Writers' Conference. His poems have appeared in numerous magazines including *The New Yorker*, *Antaeus*, *Atlantic Monthly*, *Poetry*, and *The Paris Review*. His first collection of poems, *Mysteries of the Horizon*, was published in 1972. For the past three years Mr. Raab has held a fellowship from the University of Michigan Society of Fellows.

DATE DUE
